Zorro!

蒙面俠梭羅

U0132442

商務印書館

This Chinese edition of *Zorro!*
has been published with the written permission of
Black Cat Publishing.

The copyright of this Chinese edition is owned by
The Commercial Press (H.K.) Ltd.

Name of Book: Zorro!
Retold by: Sally M. Stockton
Editors: Monika Marszewska, Elvira Poggi Repetto
Design: Nadia Maestri
Illustrations: Alfredo Belli
Edition: ©1999 Black Cat Publishing
                  an imprint of Cideb Editrice, Genoa, Canterbury

系 列 名：Black Cat 優質英語階梯閱讀 · Level 1
書　　名：蒙面俠梭羅
顧　　問：Angeli Lau
責任編輯：傅　伊
封面設計：張　毅
出　　版：商務印書館（香港）有限公司
　　　　　香港筲箕灣耀興道 3 號東滙廣場 8 樓
　　　　　http://www.commercialpress.com.hk
發　　行：香港聯合書刊物流有限公司
　　　　　香港新界大埔汀麗路 36 號中華商務印刷大廈 3 字樓
印　　刷：中華商務彩色印刷有限公司
　　　　　香港新界大埔汀麗路 36 號中華商務印刷大廈
版　　次：2012 年 4 月第 10 次印刷
　　　　　© 商務印書館（香港）有限公司
　　　　　ISBN 978 962 07 1634 8
　　　　　Printed in Hong Kong

出版説明

　　本館一向倡導優質閱讀，近年來連續推出了以"Q"為標識的 "Quality English Learning 優質英語學習"系列，其中《讀名著學英語》叢書，更是香港書展入選好書，讀者反響令人鼓舞。推動社會閱讀風氣，推動英語經典閱讀，藉閱讀拓廣世界視野，提高英語水平，已經成為一種潮流。

　　然良好閱讀習慣的養成非一日之功，大多數初、中級程度的讀者，常視直接閱讀厚重的原著為畏途。如何給年輕的讀者提供切實的指引和幫助，如何既提供優質的學習素材，又提供名師的教學方法，是當下社會關注的重要問題。針對這種情況，本館特別延請香港名校名師，根據多年豐富的教學經驗，精選海外適合初、中級英語程度讀者的優質經典讀物，有系統地出版了這套叢書，名為《Black Cat 優質英語階梯閱讀》。

　　《Black Cat 優質英語階梯閱讀》體現了香港名校名師堅持經典學習的教學理念，以及多年行之有效的學習方法。既有經過改寫和縮寫的經典名著，又有富創意的現代作品；既有精心設計的聽、說、讀、寫綜合練習，又有豐富的歷史文化知識；既有彩色插圖、繪圖和照片，又有英美專業演員朗讀作品的 CD。適合口味不同的讀者享受閱讀之樂，欣賞經典之美。

　　《Black Cat 優質英語階梯閱讀》由淺入深，逐階提升，好像參與一個尋寶遊戲，入門並不難，但要真正尋得寶藏，需要投入，更需要堅持。只有置身其中的人，才能體味純正英語的魅力，領略得到真寶的快樂。當英語閱讀成為自己生活的一部分，英語水平的提高自然水到渠成。

商務印書館 (香港) 有限公司
編輯部

使用説明

1 應該怎樣選書？

按閱讀興趣選書

《Black Cat 優質英語階梯閱讀》精選世界經典作品，也包括富於創意的現代作品；既有膾炙人口的小説、戲劇，又有非小説類的文化知識讀物，品種豐富，內容多樣，適合口味不同的讀者挑選自己感興趣的書，享受閱讀的樂趣。

按英語程度選書

《Black Cat 優質英語階梯閱讀》現設 Level 1 至 Level 6，由淺入深，涵蓋初、中級英語程度。讀物分級採用了國際上通用的劃分標準，主要以詞彙（vocabulary）和結構（structures）劃分。

Level 1 至 Level 3 出現的詞彙較淺顯，相對深的核心詞彙均配上中文解釋，節省讀者查找詞典的時間，以專心理解正文內容。在註釋的幫助下，讀者若能流暢地閱讀正文內容，就不用擔心這一本書程度過深。

Level 1 至 Level 3 出現的動詞時態形式和句子結構比較簡單。動詞時態形式以現在時（present simple）、現在時進行式（present continuous）、過去時（past simple）為主，句子結構大部分是簡單句（simple sentences）。此外，還包括比較級和最高級（comparative and superlative forms）、可數和不可數名詞（countable and uncountable nouns）以及冠詞（articles）等語法知識點。

Level 4 至 Level 6 出現的動詞時態形式，以現在完成時（present perfect）、現在完成時進行式（present perfect continuous）、過去完成時（past perfect continuous）為主，句子結構大部分是複合句（compound sentences）、條件從句（1st and 2nd conditional sentences）等。此外，還包括情態動詞（modal verbs）、被動形式（passive forms）、動名詞（gerunds）、

短語動詞（phrasal verbs）等語法知識點。

　　根據上述的語法範圍，讀者可按自己實際的英語水平，如詞彙量、語法知識、理解能力、閱讀能力等自主選擇，不再受制於學校年級劃分或學歷高低的約束，完全根據個人需要選擇合適的讀物。

❷ 怎樣提高閱讀效果？

　　閱讀的方法主要有兩種：一是泛讀，二是精讀。兩者各有功能，適當地結合使用，相輔相成，有事半功倍之效。

　　泛讀，指閱讀大量適合自己程度（可稍淺，但不能過深）、不同內容、風格、體裁的讀物，但求明白內容大意，不用花費太多時間鑽研細節，主要作用是多接觸英語，減輕對它的生疏感，鞏固以前所學過的英語，讓腦子在潛意識中吸收詞彙用法、語法結構等。

　　精讀，指小心認真地閱讀內容精彩、組織有條理、遣詞造句又正確的作品，着重點在於理解 "準確" 及 "深入"，欣賞其精彩獨到之處。精讀時，可充分利用書中精心設計的練習，學習掌握有用的英語詞彙和語法知識。精讀後，可再花十分鐘朗讀其中一小段有趣的文字，邊唸邊細心領會文字的結構和意思。

　　《Black Cat 優質英語階梯閱讀》中的作品均值得精讀，如時間有限，不妨嘗試每兩個星期泛讀一本，輔以每星期挑選書中一章精彩的文字精讀。要學好英語，持之以恆地泛讀和精讀英文是最有效的方法。

❸ 本系列的練習與測試有何功能？

　　《Black Cat 優質英語階梯閱讀》特別注重練習的設計，為讀者考慮周到，切合實用需求，學習功能強。每章後均配有訓練聽、說、讀、寫四項技能的練習，分量、難度恰到好處。

聽力練習分兩類，一是重聽故事回答問題，二是聆聽主角對話、書信朗讀、或模擬記者訪問後寫出答案，旨在以生活化的練習形式逐步提高聽力。每本書均配有 CD 提供作品朗讀，朗讀者都是專業演員，英國作品由英國演員錄音，美國作品由美國演員錄音，務求增加聆聽的真實感和感染力。多聆聽英式和美式英語兩種發音，可讓讀者熟悉二者的差異，逐漸培養分辨英美發音的能力，提高聆聽理解的準確度。此外，模仿錄音朗讀故事或模仿主人翁在戲劇中的對白，都是訓練口語能力的好方法。

閱讀理解練習形式多樣化，有縱橫字謎、配對、填空、字句重組等等，注重訓練讀者的理解、推敲和聯想等多種閱讀技能。

寫作練習尤具新意，教讀者使用網式圖示（spidergrams）記錄重點，採用問答、書信、電報、記者採訪等多樣化形式，鼓勵讀者動手寫作。

書後更設有升級測試（Exit Test）及答案，供讀者檢查學習效果。充分利用書中的練習和測試，可全面提升聽、說、讀、寫四項技能。

④ 本系列還能提供甚麼幫助？

《Black Cat 優質英語階梯閱讀》提倡豐富多元的現代閱讀，巧用書中提供的資訊，有助於提升英語理解力，擴闊視野。

每本書都設有專章介紹相關的歷史文化知識，經典名著更有作者生平、社會背景等資訊。書內富有表現力的彩色插圖、繪圖和照片，使閱讀充滿趣味，部分加上如何解讀古典名畫的指導，增長見識。有的書還提供一些與主題相關的網址，比如關於不同國家的節慶源流的網址，讓讀者多利用網上資源增進知識。

Contents

The story is recorded in full. 故事錄音

 These symbols indicate the beginning and end of the extracts linked to the listening activities.   聽力練習開始和結束的標記

Introduction

This story takes place in Spanish California during the first part of the 1800's. The word "Zorro" means fox [1] in Spanish. The story of Zorro is an old legend from Spanish California.

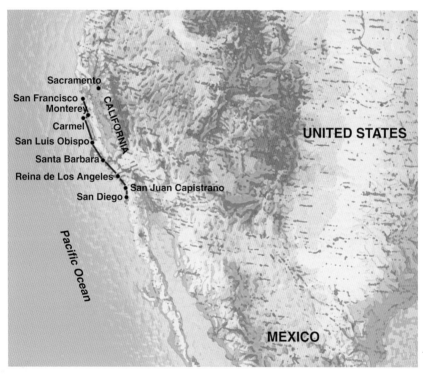

El Camino Real.

---

1. fox :

Here are some words from the story. Many of them are Spanish.

presidio
要塞

friar
修士

patio
庭院

señor
先生

señorita
小姐

sombrero
闊邊帽

sword
劍

hacienda
大莊園

CHAPTER ONE

A Stormy Night

Reina de Los Angeles *(see map on page 9)* is a Spanish village in southern California.

In the village there is a military [1] presidio with Spanish soldiers and their horses. There is also a Spanish church called a mission. The Spanish friars live here. All around the village there are big homes with patios.

Tonight there is a terrible storm and it is raining. Inside the village tavern [2] there are soldiers and other men. They are eating and drinking. Sergeant [3] Pedro Gonzales is at the tavern. He is a big, strong man.

"What a horrible night! It always rains in February. Where is Zorro on this stormy night?" asks one soldier.

"Zorro! Don't say that name! He is a bandit [4] and a criminal," says Sergeant Gonzales.

---

1. **military** : 軍人的，軍事的。
2. **tavern** : 小酒店。
3. **sergeant** : 中士。
4. **bandit** : 土匪，強盜。

Zorro!

"He is the terror [1] of southern California," says another soldier.

"People say that he takes from the rich and gives to the poor. He is a friend of the natives [2] and the friars. He punishes dishonest people," says an old man.

"Ha! Zorro is a big mystery. Who is he? Where is he from? He wears a black mask and no one can see his face. He travels on the El Camino Real [3] on his fast horse. He is very good with his sword," says the Sergeant.

"Yes, and he leaves his mark—the Z—everywhere," says the old man.

"No one can stop him. The Governor [4] of California offers a big reward [5] for the capture [6] of Zorro," says one soldier.

1.	**terror**：恐怖。	4.	**Governor**：總督。
2.	**natives**：當地人。	5.	**reward**：酬金。
3.	**El Camino Real**：舊加利福尼亞內的一條長路名（見第**9**頁）。	6.	**capture**：逮捕。

A Stormy Night

At that moment a man enters the tavern. He is young and handsome. He has black hair and dark eyes. He has fine clothes.

"Don [1] Diego Vega, my friend!" says Sergeant Gonzales. "Your clothes are wet. Why are you out on this rainy night?"

Don Diego smiles and says, "I am going home, but I am cold and wet. I want something to drink."

"Come and stand near the fire," says the Sergeant. "Here is a glass of wine."

"Thank you, my friend," says Don Diego.

"We are talking about Zorro. Everyone is scared [2] of him, but I am not! I am ready to fight Zorro and win! I am a champion [3] with the sword. What do you think, Don Diego?" asks the Sergeant. "Everyone talks about this mysterious [4] man with a mask. Many people say good things about him," says Don Diego.

"I want to fight him and capture him! I want the big reward," says Sergeant Gonzales.

"No, no! Don't talk about fighting. I hate fighting and I hate violence [5]. I think Zorro is sincere. He punishes only bad people. He protects the poor, the natives and the friars. Let Zorro do his work," says Don Diego.

"You are a kind man. You like music and poetry. You don't understand, my friend. You are rich and noble," says the Sergeant.

Don Diego smiles and says, "It's 6 p.m. I must return to my hacienda. Good night everyone." He opens the tavern door and goes out into the rain.

---

1. **Don**：（西班牙語，用於人名前的尊稱）先生。
2. **scared**：感到害怕的。
3. **champion**：優勝者。
4. **mysterious**：神秘的。
5. **violence**：暴行。

UNDERSTANDING THE TEXT

1 Read Chapter 1 again and fill in the gaps. Use the words in the mask to help you.

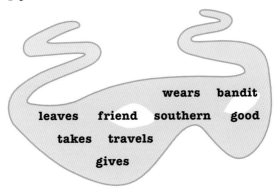

wears   bandit
leaves   friend   southern   good
takes   travels
gives

a. Zorro is a .... and a criminal.

b. He is the terror of .... California.

c. He .... from the rich and .... to the poor.

d. He is a .... of the natives and the friars.

e. He .... a black mask.

f. He .... on the El Camino Real.

g. He .... his mark—the Z—everywhere.

2 Answer these questions about Don Diego Vega.

a. Is he an old man?

b. What does he look like?

c. What does he wear?

d. What does he like?

e. Does he like Zorro? Why?

f. Is he poor?

3 **What are their nationalities?**
Complete the table below. You can use a dictionary to help you.

COUNTRY	NATIONALITY
Spain	Spanish
France	....
....	Italian
U.S.A.	....
Germany	....
....	English

Look at the example and write similar sentences.
Carmen is from Madrid. She is Spanish.

a. Paolo lives in Rome.
He ....

b. Catherine is from Paris.
She ....

c. John and Jane live in New York.
....

d. Elizabeth and I come from Liverpool.
....

e. You and your friends are from Berlin.
....

4. What's the weather like?
Match the weather words with the pictures.

stormy

foggy

windy

sunny

snowy

cloudy

rainy

Look at the example and then complete the sentences.

Stay at home! It's a stormy night.

a. I can't see the road! ....

b. Put your sunglasses on! ....

c. Where are my skis? ....

d. Let's fly our kites! ....

e. Take your umbrella! It's a .... day.

CHAPTER TWO

A Surprise Visit

At 7 p.m. the door of the tavern opens. Sergeant Gonzales and the other soldiers are standing near the fire. They are talking about Don Diego Vega and his rich family.

They turn around and see a man. He is dressed in black. He has a black mask and a black sombrero. It is Zorro!

"Good evening! My name is Zorro!" says the masked man.

The men in the tavern are very surprised and scared.

Sergeant Gonzales looks at him carefully and says, "What do you want, bandit?"

Zorro laughs loudly. He looks at Sergeant Gonzales and says, "I am here to punish you, Sergeant!"

"What do you mean?" says the Sergeant.

"You beat [1] the poor natives. I am a friend of the natives. I am here to punish you."

---

1. beat :

 Zorro!

"You idiot [1]! The governor wants you dead or alive!" says Sergeant Gonzales. He pulls out his sword and wants to fight.

At that moment Zorro pulls out a pistol [2]. He watches everyone carefully. Sergeant Gonzales looks at the pistol and says, "Courageous [3] men don't use pistols. They use swords. Perhaps you are not courageous, Zorro?"

"This pistol is necessary because you have many friends in this tavern. Everyone must go near the fire and stay there!" says Zorro. "Now I have a pistol in my left hand and a sword in my right hand. I am ready to punish you."

"Fight, señor!" says the Sergeant.

They start fighting. Their swords move quickly and make a lot of noise. The two men are good fighters. Zorro is fast and light. Sergeant Gonzales is slow and heavy. Zorro jumps onto a table. Then he jumps onto a chair. The fight continues and Sergeant Gonzales' sword falls onto the floor. His face is white. He is scared. Zorro slaps [4] his face and says, "This is your punishment." Then he makes a Z on the Sergeant's shirt with his sword.

Zorro runs to the window. He opens it and says, "Good evening, gentlemen!" He jumps out of the window and disappears.

---

1. **idiot** : 笨蛋。
2. **pistol** : 手槍。
3. **courageous** : 勇敢的。
4. **slaps** : 打耳光。

UNDERSTANDING THE TEXT

1 What does Zorro do?
Find the missing verbs in Chapter 2.

a. .... the door.

b. .... loudly.

c. .... at Sergeant Gonzales.

d. .... out a pistol.

e. .... everyone carefully.

f. .... onto a table.

g. .... the Sergeant's face.

h. .... a Z.

i. .... to the window.

j. .... out of the window.

2 Revise your grammar!
In the present simple（現在時）the third person singular（第三人稱單數）endings are *S*, *ES* and *IES*.
Change the following verbs and put them into the correct columns.

travel   rain   wash   play   go   work
start   study   watch   cry   do

S	ES	IES

Fill in the blanks with the correct form of the verbs given in brackets.

a. Dad .... his car on Sundays.   (wash)

b. Zorro .... on a horse.   (travel)

c. It .... a lot in England.   (rain)

d. Peter .... to the gym twice a week.   (go)

e. Frank .... in the afternoon.   (study)

f. Sergeant Gonzales .... the poor.   (beat)

g. Alice never .... TV.   (watch)

3 Give the opposites of these words.

a. rich   ....

b. alive   ....

c. left   ....

d. bad   ....

e. heavy   ....

f. fast   ....

4 What's the time?

a. 7:10   It's ten past seven.

b. 9:15   ....

c. 6:25   ....

d. 8:30   ....

e. 1:40   ....

f. 3:45   ....

g. 4:55   ....

CHAPTER THREE

The Pulido Hacienda

The next day is warm and sunny. Don Diego gets on his beautiful horse. He rides to Don Carlos Pulido's big hacienda.

Don Carlos is a good friend of Don Diego's family. Both families are rich and important. But the Governor does not like Don Carlos. He creates problems for him. He wants to take Don Carlos' land.

Don Carlos is happy to see his friend. "Good morning, Don Diego. What a nice surprise! Come and sit in the patio."

"Thank you. I am here to say something very important," says Don Diego. "I am almost 25 years old. My father wants me to get married and start a family. I am not interested in marriage. I think love and marriage are boring [1] but I must obey [2] my father. How old is your daughter Lolita?" asks Don Diego.

---

1. **boring** : 無趣的。
2. **obey** : 服從。

The Pulido Hacienda

"Lolita is 18 years old and she is very beautiful," answers Don Carlos.

"You have a fine family. With your permission [1], I want to marry your daughter," says Don Diego.

Don Carlos smiles and is happy. "This is an honour for our family. You have my permission! Do you want to see Lolita?"

"I think I must!" answers Don Diego.

Don Carlos calls her and she comes to the patio. Lolita is a lovely girl with long black hair and dark eyes.

"Good morning, señorita. There is something I must tell you," says Don Diego smiling. "I want to marry you and your father approves."

"Oh, señor!" exclaims [2] Lolita. "You want to marry me!" She is surprised and her face is red.

1. **permission** : 准許。   2. **exclaims** : 大叫。

"Think about it today. One of my servants can bring me your answer tomorrow."

"But why can't you come tomorrow?" asks Lolita.

"Oh, your hacienda is far. I get tired when I ride my horse. I prefer [1] to stay at home and rest."

"What! You want to marry me and you don't want to visit me! Is this your idea of love? I want to marry a strong, romantic man. You are young and rich, but you are not strong or romantic! Do you have a heart?" says Lolita. She is angry. She runs away and tells her mother.

Doña Catalina says, "You are lucky, Lolita. Don Diego is very rich. He comes from a noble family. The Governor likes his family. This is a big opportunity [2] for our family."

In the afternoon Lolita is alone in the patio. She is thinking about Don Diego. Suddenly she hears a noise and turns around. She sees Zorro standing in front of her. "Zorro!" she whispers [3].

"Don't be afraid, señorita. I only punish corrupt [4] people. I like your father because he is honest. I am here to admire [5] your beauty."

"What! You must go away. You are in great danger," says Lolita.

"You are beautiful and kind, Lolita," says Zorro. "Let me kiss your hand." Zorro takes her small hand and kisses it. Lolita looks into his eyes and smiles. Then she runs into the house. "What a courageous man! He is a bandit but I like him," Lolita thinks.

---

1. **prefer**：（在相比較之下）喜歡某事物或某人。
2. **opportunity**：機會。
3. **whispers**：小聲說。

4. **corrupt**：腐敗的。
5. **admire**：欣賞，愛慕。

UNDERSTANDING THE TEXT

 Fill in the gaps with the correct prepositions from the sombrero. Some words can be used more than once.

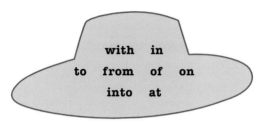

with   in

to   from   of   on

into   at

a. Don Diego gets .... his beautiful horse and rides .... Don Carlos' hacienda.

b. Don Carlos is a good friend .... Don Diego's family.

c. Lolita is a lovely girl .... long black hair.

d. Don Diego prefers to stay .... home and rest.

e. Don Diego comes .... a noble family.

f. Zorro is .... great danger.

g. Lolita looks .... Zorro's eyes, then she runs .... the house.

2 **Listen to the first three paragraphs of Chapter 3 and circle the words that you hear.**

stormy   warm   mission   home   horse

presidio   hacienda   friars

friend   family   fast   rich   sword

strong   handsome   happy

surprise   night

Complete the sentences with some of the words provided.

hair	eyes	face	hand	heart

a. Lolita's .... is red.

b. Lolita says to Don Diego: "Do you have a ....?"

c. Zorro takes Lolita's .... and kisses it.

Now find the words and circle them. Be careful: Zorro is here!

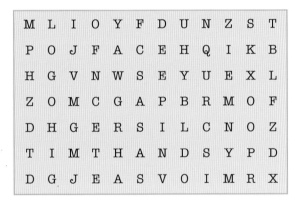

M	L	I	O	Y	F	D	U	N	Z	S	T
P	O	J	F	A	C	E	H	Q	I	K	B
H	G	V	N	W	S	E	Y	U	E	X	L
Z	O	M	C	G	A	P	B	R	M	O	F
D	H	G	E	R	S	I	L	C	N	O	Z
T	I	M	T	H	A	N	D	S	Y	P	D
D	G	J	E	A	S	V	O	I	M	R	X

 Meet Lolita!
Unscramble these sentences and put the words in the right order.

a. daughter / Don Carlos's/ is / Lolita.

b. 18 / She / old / is / years.

c. got / She / long / black / and / dark / has / eyes / hair.

d. She / romantic / marry / man / wants / to / a / strong.

e. because / Zorro / loves / She / courageous / is / man / a / he.

5 Are these sentences true (T) or false (F)? Correct the false ones.

		T	F
a.	The next day Don Diego rides to the tavern.	☐	☐
b.	The Governor likes Don Carlos.	☐	☐
c.	Don Diego isn't interested in marriage.	☐	☐
d.	Lolita is a beautiful girl.	☐	☐
e.	Don Diego never gets tired.	☐	☐
f.	Lolita is happy, so she runs away.	☐	☐
g.	Lolita likes Zorro.	☐	☐

Spanish California

Read this page from a Spanish friar's diary and learn about Spanish California.

La Purísima Concepción.

San Diego de Alcalà, April 22, 1802

California is a beautiful place. It is always warm and sunny. There are flowers and fruit trees everywhere.

I live in the San Diego de Alcalà mission. Other friars live here too. San Diego de Alcalà is the first Californian mission. Many Indians live in the village near the mission. They are friendly. We teach them many things. Spanish families live here too. Some rich families have big haciendas.

The village of San Diego de Alcalà is growing quickly. It is near the Pacific

Santa Barbara County Courthouse.

The mission of San Diego de Alcalà.

Ocean. Spanish ships often come and bring things from Spain. Some villages have military presidios. Spanish soldiers live in the presidios.

Today Friar Junipero Serra is visiting us. He is a very important friar. He travels and builds missions in California. He wants to build 21 missions! Some big missions in California are:

- *San Juan Capistrano*
- *Reina de Los Angeles*
- *Santa Barbara*
- *San Luis Obispo*
- *Carmel*
- *San Francisco*

A typical hacienda.

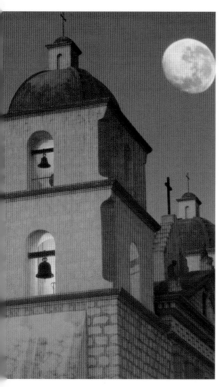

Santa Barbara mission.

San Gabriel mission.

San Carlos Borromeo mission.

The friars' bedroom.

1 **Tick ✔ the correct answer.**

a. In California it is always
- ☐ cold and sunny
- ☐ warm and sunny
- ☐ warm and rainy

b. San Diego de Alcalà is
- ☐ a presidio
- ☐ a mission
- ☐ a hacienda

c. Many Indians live
- ☐ in the mission
- ☐ in the village near the mission
- ☐ in reservations

d. The Indians are
- ☐ friendly
- ☐ dangerous
- ☐ rich

e. Some rich families have
- ☐ big ships
- ☐ big villages
- ☐ big haciendas

f. San Diego de Alcalà is near
- ☐ the Mississippi River
- ☐ the Pacific Ocean
- ☐ the Atlantic Ocean

g. Ships come and bring things from
- ☐ England
- ☐ France
- ☐ Spain

h. Some villages have
- ☐ old castles
- ☐ military presidios
- ☐ big farms

CHAPTER FOUR

Captain Ramón

At 8 p.m. it is dinner time at Don Carlos' home. The family is sitting at the table. Someone knocks at the door. A servant opens it and Zorro appears!

Don Carlos, his wife and daughter stand up. They are scared.

"Good evening," says Zorro. "Don't be scared! You are an honest man, Don Carlos. I only want some food and drink."

Zorro goes to Lolita and whispers, "I cannot forget this afternoon in the patio."

"You must not come here. It is dangerous!" whispers Lolita.

Suddenly a young Spanish soldier enters the house. He is Captain Ramón. He wants to arrest [1] Zorro. Zorro takes out his sword and they fight. The two men are good fighters.

"I want to arrest you, Zorro," says Captain Ramón. "You are a public enemy [2]."

---

1. **arrest**：逮捕。
2. **enemy**：敵人。

Zorro!

"You cannot arrest me!" says Zorro.

Zorro injures [1] the Captain's shoulder [2] with his sword. The Captain falls to the floor.

"Please help the Captain!" Zorro says to Don Carlos. Then he smiles at Lolita and rides away on his horse.

Don Carlos' wife and daughter look after Captain Ramón's injured shoulder. The Captain likes beautiful Lolita.

1. **injures**：傷害。
2. **shoulder**：

Captain Ramón

"Don Carlos," says the Captain, "I like Lolita very much. I come from a good family and I am the Governor's friend. I am 23 years old and I am the Captain of the Presidio. May I court [1] Lolita?"

"First, I must explain something," says Don Carlos. "Don Diego Vega wants to court Lolita too. Lolita must choose her husband herself, but you have my permission to court her!"

The next morning there is a lot of noise at the Presidio. Don

1. **court** : 追求。

 Zorro!

Diego and other men are watching. There are many soldiers on their horses. They are listening to Sergeant Gonzales. "Today is an important day. We must find Zorro! We must look in every hacienda and in every home. Remember the Governor's big reward. Let's go!"

That morning Don Diego sends a letter to Don Carlos:

My dear friend,

Sergeant Gonzales and his soldiers want to arrest Zorro. It is dangerous for you and your family to stay at your hacienda. Please come to my home in Reina de Los Angeles. It is safe in my home. I must go away for a few days.

Your friend,
Diego Vega

Don Carlos receives the letter and says, "What a generous [1] invitation! Don Diego wants to protect Lolita. We must accept the invitation. Let us go immediately [2]!"

---

1. **generous**：慷慨的。　　　2. **immediately**：立刻。

UNDERSTANDING THE TEXT

 Correct the following sentences.

a. Don Carlos and his family have lunch at 8 p.m.

b. Don Carlos, his wife and son stand up.

c. A Spanish friar enters the house and wants to arrest Zorro.

d. Zorro and Captain Ramón are good players.

e. Zorro injures the Captain's leg.

f. The Captain likes Doña Catalina.

g. Don Diego reads a letter from Don Carlos.

 Who does the following actions? The names are in the sword.

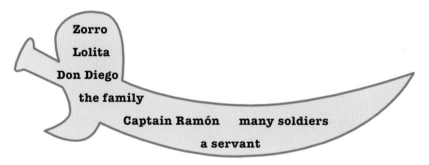

Zorro
Lolita
Don Diego
the family
Captain Ramón   many soldiers
a servant

Who...

a. is sitting at the table?   ....

b. opens the door?   ....

c. whispers to Zorro?   ....

d. takes out his sword?   ....

e. falls to the floor?   ....

f. is listening to Sergeant Gonzales?   ....

g. sends a letter?   ....

Fill in the gaps with *your, her, his, their*.

a. The Captain falls to the floor. .... shoulder is injured.

b. Zorro rides away on .... horse.

c. Lolita must choose .... husband herself.

d. There are many soldiers on .... horses.

e. "Don Carlos, you are in danger at .... hacienda!"

4 **Look at this sentence:**

Zorro and Captain Ramón fight: they are good fighters.

***Fight* is a verb. *Fighter* is a noun formed by adding *-er* to the verb. Fill in the blanks.**

VERB	NOUN
fight	fighter
teach	....
....	singer
dance	....
....	player
report	....
photograph	....
....	speaker

CHAPTER FIVE

Lolita is in Love

Don Diego's home in Reina de Los Angeles is big and beautiful. There are many rooms and many servants. There is a big garden with trees and flowers. Don Carlos and his family like it.

"Lolita! Marry Don Diego and this beautiful home is yours!" says Doña Catalina.

"I don't love Don Diego! I don't want to marry him!" says Lolita.

"Lolita, tonight your mother and I must visit our old friends," says Don Carlos. "Can you stay at home alone?"

"Yes, of course. There are many books in Don Diego's library. I can stay here and read."

Lolita looks at the books. "How strange!" she thinks. "Don Diego has books about love, passion[1], adventure[2], horses and

1. **passion** : 激情。
2. **adventure** : 冒險。

Zorro!

heroes! But he isn't strong and is so lifeless [1]!"

Suddenly there is a knock at the door. A servant opens it and Captain Ramón enters. He goes to the library and sees Lolita.

"I'm glad you are alone! You are a very beautiful girl. I have your father's permission to court you! I know that Don Diego wants to marry you, but I am superior [2] in every way. Don Diego has no courage. He is a weak man. I am the Captain of the Presidio."

"I am alone," says Lolita. "You cannot stay here. It is not right. Please go away!"

Captain Ramón takes her hand and says, "Don't run away, Lolita. You are mine! Kiss me!"

Lolita pushes him away and says, "I don't want to kiss you. Go away!"

"No, you are mine. You must marry me. Now kiss me!"

Lolita slaps the Captain's face.

At that moment Zorro appears in the library and says, "Captain Ramón, you are a villain [3]. Don't touch Lolita! Leave this house immediately!"

"I cannot forget this terrible insult [4]," says the Captain. His face is red with anger. Zorro opens the door and kicks him out.

"Thank you for your help, Zorro. You are courageous and kind. I love you!" Lolita says.

"Dear Lolita, I love you too!" says Zorro.

They embrace [5] and kiss.

---

1. **lifeless**：沒有生氣的。
2. **superior**：優越的。
3. **villain**：壞人，惡棍。
4. **insult**：侮辱。
5. **embrace**：擁抱。

UNDERSTANDING THE TEXT

1 **Here's a short summary of Chapter 5. Fill in the gaps with the words given.**

> door   library   home   hand   big
>
> villain   out   embrace   slaps
>
> enters   kiss   appears

Don Carlos and his family are at Don Diego's ....

It is a .... and beautiful house.

Lolita is in the .... when Captain Ramón
.... He takes Lolita's .... and wants to
.... her, but Lolita .... his face. At that
moment Zorro .... and tells Captain Ramón he is a
.... Then he opens the .... and kicks him
....

Lolita and Zorro .... and kiss.

2 **Unscramble these words from Chapter 5. Then find and circle them in the word square.**

a. L V L I A N I   ....

b. S R W F E L O   ....

c. O R O D   ....

d. B L Y R R A I   ....

e. N R G E D A   ....

f. S E R T E   ....

g. N A H D   ....

h. E C A F   ....

F	B	V	I	L	L	A	I	N
A	L	P	Z	M	I	C	E	G
C	H	O	J	N	B	P	S	Q
E	D	T	W	U	R	X	W	H
Z	O	H	V	E	A	P	N	A
O	O	G	A	R	D	E	N	
E	R	Y	I	U	Y	S	T	D
Q	T	R	E	E	S	O	U	M

CHAPTER SIX

Friar Felipe

Captain Ramón returns to the Presidio. He is furious [1].

"I must punish Lolita, her family and Zorro for their insults!" he thinks. "I must write a letter to the Governor. I must tell him that Don Carlos Pulido and his family are traitors [2]. They are Zorro's friends. They help him."

He writes the letter and sends it to the Governor. He smiles and says, "I want to see the Pulido family in prison!"

"I want to see *you* in prison!" says a man's voice. Captain Ramón turns around and sees Zorro. "You are a villain. Fight me but don't hurt the Pulido family!" says Zorro.

"Sergeant Gonzales, come quickly!" says the Captain. "Zorro is here."

Zorro disappears.

1. **furious**：憤怒的。
2. **traitors**：叛徒，賣國賊。

orro!

"I am here, Captain," says Sergeant Gonzales.

"Take all the soldiers and find Zorro! We must capture him."

The soldiers follow Zorro. The night is dark. It is difficult to follow him because his horse is very fast. The next morning the soldiers return to the Presidio. They are tired and angry. Zorro is still free.

There are many people in front of the Presidio that morning. Don Diego is there too. Something is happening.

An old friar is standing before the magistrate [1]. The old friar is in chains [2]. "I am not a thief [3]," says the old friar. "I am a poor friar."

"What is happening?" asks Don Diego.

"This old friar is a thief. He must be punished," answers the magistrate.

"That is impossible. Friar Felipe is an honest man. I know him," says Don Diego.

"No, you are wrong," says the cruel magistrate.

He calls two soldiers and says, "Whip [4] this friar 15 times."

The soldiers whip the old friar and he falls to the ground. Don Diego is very angry because Friar Felipe is his friend. He returns to his father's hacienda.

"Good afternoon, my son," says his father, Don Alejandro Vega. "I am happy to see you. Come and tell me about Lolita. Does she want to marry you?"

---

1. **magistrate** : 地方法官。

2. **chains** :

3. **thief** : 賊，小偷。

4. **whip** :

 Zorro!

"I like Lolita, but she doesn't like me. She likes romantic [1] men. What can I do?" asks Don Diego.

"Girls like courageous, romantic men. You must talk about love. You must play the guitar [2] and sing love songs. Give her some flowers. This is what young men do. Wake up, Diego!" says Don Alejandro.

"How ridiculous [3]! I cannot do these stupid things," says Don Diego.

"You must try. Lolita is a lovely girl," says Don Alejandro.

"There are many problems in my life. I want to rest and meditate [4]," says Don Diego.

1. **romantic**：浪漫的。

2. **guitar**：

3. **ridiculous**：可笑的，荒謬的。

4. **meditate**：沉思。

UNDERSTANDING THE TEXT

1 Answer the questions.

a. Where does Captain Ramón return to?

b. What does he do?

c. Why is it difficult to follow Zorro?

d. How do the soldiers feel the next morning?

e. Who is Felipe?

f. Where does Don Diego return to?

2 Read the passage and try to fill in the missing words. Now listen and check if you were right.

There are many .... in front of the Presidio that .... Don Diego is there too. Something is happening.

An old .... is standing before the magistrate. The old friar is in .... "I am not a thief," says the old friar. "I am a .... friar."

"What is happening?" asks Don Diego.

"This old friar is a thief. He must be punished," answers the ....

"That is impossible. Friar Felipe is an honest man. I know him," says Don Diego.

"No, you are wrong," says the cruel magistrate.

He calls .... soldiers and says, "Whip this friar 15 times."

The soldiers .... the old friar and he falls to the ground. Don Diego is very angry because Friar Felipe is his .... He returns to his ....'s hacienda.

CHAPTER SEVEN

The Avengers[1]

In the evening the magistrate and his friends are in the tavern. They are laughing about the old friar.

"Why are you laughing?" asks a mysterious voice. They look at the door of the tavern and see Zorro. He has a pistol in one hand and a sword in the other.

"Magistrate! I am here to punish you," says Zorro. "Old Friar Felipe is not a thief and you know it."

"I am an important magistrate. I don't like friars because they are your friends, Zorro."

Zorro gives a whip to the magistrate's friend and says, "Now whip this corrupt magistrate 15 times."

"But I cannot do this," says the friend.

"Whip him or I whip you!" says Zorro.

The friend whips the magistrate. After the punishment the magistrate falls to the ground.

---

1. **Avengers**：復仇者（梭羅的朋友，喜歡懲罰壞人）。

The Avengers

"This is how I punish dishonest [1] people," says Zorro.

The next day everyone talks about the magistrate's punishment. A group of young men want to help Captain Ramón arrest Zorro. They look for him in the hills and in the valleys. In the evening they go to the Vega hacienda. When Don Alejandro sees them he says, "Why are you all here?"

"We are looking for Zorro. We want to capture the bandit and get the reward. But now we are tired and hungry. Can you give us some food?"

"Yes, of course. Please come in. You can put your swords and pistols near the door," says Don Alejandro. "Sit down! Eat these cakes and drink this wine."

Don Alejandro and Don Diego talk to the young men.

At 9 p.m. Don Diego says, "Please excuse me. I am tired and I am going to bed."

"Look, Diego, these young men are not tired. It's only 9 p.m. You are young but you are always tired," says Don Alejandro.

"Yes, father, you are right. Good night everyone!"

The others eat, drink and sing.

At midnight a masked man appears at the door.

"Look, it is Zorro!" says one of the young men.

"Zorro, the bandit," they all say.

"Yes, I am Zorro, but I am not a bandit. I have principles [2] and

1. **dishonest** : 不誠實的，不正直的。
2. **principles** : 原則。

$\mathcal{Z}$orro!

I fight for them. In California we have corrupt political men, cruel magistrates and dishonest people. I want to change this. I fight to help the poor, the natives and the friars. What are your principles?"

"We want to help the poor, the natives and the friars too," says one young man.

"Our principles are the same," says another.

"Come with me and we can fight together! We can make California a better place to live," says Zorro.

The Avengers

"But who are you? Where do you live?" asks a young man.

"I can't tell you. It's a secret," says Zorro.

The young men talk together.

"Yes, we want to fight with you. We want justice [1] in California. Our new name is The Avengers!" says a young man.

"Yes, we are The Avengers," they all say.

"Good! We now fight together!" says Zorro and leaves.

---

1. **justice**：公道，正義。

UNDERSTANDING THE TEXT

1 **Put the events in chronological order. The first one is done for you.**

a. ☐ The friend whips the magistrate.

b. ☐ At midnight a masked man appears at the door.

c. ☐ They look at the door of the tavern and see Zorro.

d. ☐ The next day a group of young men look for Zorro in the hills and in the valleys.

e. ☐1 In the evening the magistrate and his friends are in the tavern.

f. ☐ The Avengers want to fight with Zorro for justice in California.

g. ☐ In the evening they go to the Vega hacienda.

h. ☐ Zorro gives a whip to the magistrate's friend.

2 **In the text find the opposite of:**

a. honest ....

b. old ....

c. wrong ....

d. disappears ....

e. alone ....

Have fun with this crossword!

ACROSS

1.

2.

3. 12.00 at night.

4. The magistrate and his friends are in the ....

5.

6.

DOWN

7.

8. The friend whips the .... .

9.

10. They punish bad people.

11.

California Today

California is an important state of the United States. About 34 million people live there. Many old Spanish missions are now important cities and towns.

Many people in California speak Spanish. Towns, streets, parks, schools, and other places often have Spanish names. The capital of California is Sacramento.

California is famous for its natural beauty. There are tall mountains, green valleys, forests, sandy beaches [1] and hot deserts.

Death Valley.

Some beautiful national parks are Yosemite, Sequoia, King's Canyon, Muir Woods, Shasta, Death Valley and Humboldt Redwoods.

Young people like Disneyland and the Great America Theme Park. Millions of tourists visit California every year. There are a lot of interesting places to visit.

San Francisco, Golden Gate Bridge.

1. **sandy beaches** : 沙灘。

The University of California, Berkeley, and Stanford University are two excellent universities. California is a cosmopolitan [1] state. People from all over the world live there.

1. **cosmopolitan** : 國際性的。

Yosemite National Park.   Disneyland.

San Francisco, UnderWater World.

Fill in the gaps with the following words in the box.

> **capital   people   young   speak   state**
> **universities   missions   parks**
> **cosmopolitan   natural   names**

California is an important .... of the United States with a population of 34 million .... Many old Spanish .... are now important cities and towns.

Many people .... Spanish and many places often have Spanish .... The .... of California is Sacramento.

California is famous for its .... beauty and its national ....

.... people like Disneyland and the Great America Theme Park.

In California there are two excellent ....

California is a .... state.

Use the letters below and write the names of:

a. two important cities ....

b. a national park ....

c. a university ....

```
A A A    B    C C    E E E E E E    F    G
I I    K    L L L    M    N N N    O O O
R R    S S S S    T    Y Y
```

CHAPTER EIGHT

The Escape[1]

"The Governor of California is here today," says Sergeant Gonzales.

"Good!" says Captain Ramón. "I must speak to him."

"Good morning, Captain Ramón," says the Governor. "I have your letter. Thank you for the information about the Pulido family. They are Zorro's friends. They are traitors. We must put them in prison and kill them! They are dangerous."

"What a good idea! My soldiers can arrest them today," says the Captain.

A group of soldiers goes to the Pulido hacienda. They arrest Don Carlos, Doña Catalina and Lolita. They take them to prison. Don Carlos is furious. His wife and daughter cry.

When Don Diego hears about this, he goes to the Governor. He asks him, "Why is the Pulido family in prison?"

The Governor answers, "They are Zorro's friends. They help and protect him. They are traitors!"

---

1. **escape** : 逃脱。

Zorro!

"I cannot believe this. I know them. They are honest people. They don't help bandits," says Don Diego.

"You are wrong, Don Diego. They must be punished. The punishment for traitors is death," says the Governor.

Early in the evening Zorro sends a message to The Avengers. The message says:

> *"Meet me at midnight at the lake.*
> *Bring your swords and pistols.*
> *Pass the word to everyone."*

At midnight Zorro and The Avengers meet at the lake. Every Avenger has a mask on his face. Zorro says, "We are here to rescue [1] Don Pulido and his family. They must escape from prison. They are innocent [2]. We must be silent, enter the prison and help them. Francisco, you take Don Carlos to the village of Pala. José, you take Doña Catalina to the Vega hacienda. I want to take Lolita to Friar Felipe. They must hide for a few days."

Zorro and The Avengers rescue the Pulido family. Don Carlos and Doña Catalina reach their destination [3]. The soldiers follow Zorro and Lolita. Zorro's horse is very fast.

Zorro and Lolita arrive at Friar Felipe's hacienda. Zorro says, "Can Lolita stay here with you for a few days? She is in danger."

"Yes, I can protect her," says Friar Felipe. Zorro kisses Lolita and says, "Always remember that I love you." Then he rides away on his horse.

---

1. **rescue**：搭救。
2. **innocent**：無辜的，清白的。
3. **destination**：目的地。

UNDERSTANDING THE TEXT

 Circle the correct answers.

a. The Governor says, "The Pulido family are Zorro's friends. They are *traitors / bandits*."

b. A group of *friends / soldiers* goes to the Pulido hacienda.

c. Don Diego says, "The Pulido family are *dangerous / honest* people. They don't help bandits."

d. "The punishment for traitors is *death / prison*," says the Governor.

e. At midnight Zorro and The Avengers meet at the *presidio / lake*.

f. Every Avenger has a *mask / sword* on his face.

g. Zorro says, "We are here to *arrest / rescue* Don Pulido and his family."

h. Zorro and Lolita arrive at Friar Felipe's *cottage / hacienda*.

 Decode Zorro's message.

7 1 9 9   10 4 3   12 6 8 2   10 6

3 11 3 8 13 6 5 3

code:

1 = A	6 = O	10 = T
2 = D	7 = P	11 = V
3 = E	8 = R	12 = W
4 = H	9 = S	13 = Y
5 = N		

— — — —   — — —   — — — —   — —   — — — — — — —.

CHAPTER NINE

The Man behind the Mask

he soldiers follow Zorro all night. They cannot capture him.

The next night Zorro silently enters the Governor's home. He wants to talk to the Governor and Captain Ramón. They are sitting near the fire.

Zorro enters and says, "Don't move and don't make a noise. I want to speak to you."

He has a pistol in one hand and a sword in the other.

"Zorro! Why are you here?" asks the Governor.

"You are here to die!" says Captain Ramón.

"No," answers Zorro, "I am here to bring justice. I am here to learn the truth. Governor, you want to punish the Pulido family. Why?"

"They are traitors. They are your friends, Zorro! They help

Zorro!

you. Your friends are my enemies," says the Governor.

"They are not traitors. They do not help me. They are not my friends," he says.

"Look at this letter. It accuses [1] them," says the Governor.

Zorro reads the letter and says, "It is Captain Ramón's letter. He accuses the Pulido family." Zorro looks at the Captain and says, "Captain, you are a liar [2] but I am here to punish you. Tell the Governor the truth about the Pulido family."

Zorro puts his pistol to Captain Ramón's head and says, "Tell the Governor the truth or I shoot [3]!"

Captain Ramón is silent. His face is white.

"Tell the truth, you liar," says Zorro.

"Yes, I am a liar. The letter is not true."

"This is terrible!" says the Governor. "You are a liar. You cannot be the Captain of the Presidio!"

At that moment the Captain pulls out his sword. He begins to fight Zorro. It is a long sword fight. In the end, Zorro kills Captain Ramón.

"The Captain is dead," says Zorro to the Governor.

1. **accuses**：指控。
2. **liar**：說謊的人。
3. **shoot**：

The Man behind the Mask

Outside the Governor's home there are many soldiers. They want to capture Zorro. He sees Lolita on her horse. He shouts, "Lolita, come with me. We must hide in the old tavern. This time it is very difficult to escape. There are soldiers all around us."

"I'm happy to be with you Zorro," says Lolita. "I'm not scared."

Zorro and Lolita hide inside the old tavern. The soldiers try to enter. Zorro is ready to fight.

Suddenly The Avengers come to rescue Zorro and Lolita. They explain many things to the Governor. The Governor pardons[1] Zorro.

Zorro and Lolita walk out of the old tavern. They are free. Everyone is happy and cheers[2].

The Governor says, "Now that you are free, show us your face!"

"Yes! Yes!" the people say.

Zorro takes off[3] his black mask.

"It's Don Diego Vega!" exclaims Sergeant Gonzales. Everyone is very surprised.

"My son, Don Diego! I can't believe it!" exclaims Don Alejandro.

Lolita looks at him and says, "Is this true or is it a dream? Are you really Don Diego?"

"Yes, my love. I am your Don Diego and your Zorro!" he says embracing Lolita. END

1. **pardons**：原諒。
2. **cheers**：歡呼，喝彩。
3. **take off**：脫掉。

UNDERSTANDING THE TEXT

 1 Are these sentences true (T) or false (F)? Correct the false ones.

	T	F
a. The soldiers follow Zorro. They capture him.	☐	☐
b. Zorro has a pistol and a sword.	☐	☐
c. Zorro puts his sword to Captain Ramón's head.	☐	☐
d. Captain Ramón is furious. His face is white.	☐	☐
e. It is a long sword fight. In the end, Zorro kills Captain Ramón.	☐	☐
f. Zorro and Lolita hide inside the hacienda.	☐	☐
g. The Avengers come to rescue the Governor.	☐	☐
h. Zorro and Lolita are free. Everyone is happy and cheers.	☐	☐
i. Zorro puts on his black mask.	☐	☐

 2 Listen to the following passage from Chapter 9 and circle the words you hear.

Suddenly The Avengers come to *arrest / rescue* Zorro and Lolita. They explain many things to the *Captain / Governor*. The Governor pardons Zorro.

Zorro and Lolita walk out of the *old / new* tavern. They are free. Everyone is *silent / happy* and cheers.

The Governor says, "Now that you are free, show us your *face / head*!"

"Yes! Yes!" the people say.

Zorro takes off his *brown / black* mask.

"It's Don Diego Vega!" exclaims Sergeant Gonzales. Everyone is very *surprised / angry*.

"My son, Don Diego! I can't believe it!" exclaims Don *Felipe / Alejandro*.

Lolita looks at him and says, "Is this true or is it a dream? Are you really *Don Alejandro / Diego*?"

"Yes, my love. I am your Don Diego and your Zorro!" he says embracing *Doña Catalina / Lolita*.

Are you a hero?

1. When your friends do well at school

   a. you are happy for them
   b. you want to do well too
   c. you are a bit jealous

2. When your classmates don't understand the lesson you

   a. always help them
   b. only help your best friends
   c. don't help them

. Why do you tell lies?

   a. to help a friend
   b. to protect yourself
   c. to create problems

4. When there is something new to learn, you are

   a. enthusiastic
   b. scared
   c. indifferent

Your boyfriend / girlfriend must be

   a. kind and generous
   b. very intelligent
   c. rich

6. During the summer holidays you like

  a. exciting and dangerous adventures
  b. visiting museums
  c. relaxing on the beach

7. When you see a cat in a tree

  a. you climb the tree and help the cat
  b. you call someone
  c. you do nothing

8. When you buy a present for your friend, you always get

  a. something your friend likes
  b. something you like
  c. something that costs very little.

There are 3 points for every "**a**" answer,
2 points for every "**b**" answer,
1 point for every "**c**" answer.

8-14: you must try harder!
15-19: you're not a hero yet!
20-24: you're a young hero!

EXIT TEST

CONTEXT

 Circle the correct word.

The story of Zorro takes place in ¹*Mexico/Spanish California* during the ²*first/last* part of the 1800's. The word "Zorro" means ³*cat/fox* in Spanish. There are friars and missions in Spanish California. The ⁴*friars/soldiers* teach the Indians many things. There are also ⁵*poor/rich* families with ⁶*big/small* haciendas.

COMPREHENSION

2 **Are the following sentences true (T) or false (F)? Correct the false sentences.**

	T	F
a. Zorro wears a blue mask and rides an old horse.	☐	☐
b. Don Diego is a rich young man.	☐	☐
c. Zorro makes a "Z" on Sergeant Gonzales' shirt.	☐	☐
d. Don Diego thinks marriage is romantic.	☐	☐
e. Lolita is angry with Don Diego but she likes Zorro.	☐	☐
f. Captain Ramón wants to court Lolita.	☐	☐
g. Captain Ramón wants to kiss Lolita but she slaps his face.	☐	☐
h. Friar Felipe is a dishonest man.	☐	☐
i. The Avengers want to fight with Zorro. They want to make California a better place to live.	☐	☐
j. Zorro kills Captain Ramón with his pistol.	☐	☐
k. The man behind the mask is Don Alejandro.	☐	☐

GRAMMAR

3 Put the verbs in brackets in the correct form.

a. Zorro .... onto the table. (*jump*)

b. Lolita .... to the patio to read. (*go*)

c. Captain Ramón and his men .... out of the Presidio. (*run*)

d. She .... in the library. (*study*)

e. Don Diego .... everyone carefully. (*watch*)

f. He .... the windows every week. (*wash*)

g. What .... she want? (*do*)

4 Fill in the gaps with YOUR, HIS, HER, THEIR.

a. Friar Felipe falls on the floor and hurts .... hand.

b. Lolita loves .... black horse.

c. The soldiers stop and have .... dinner.

d. "Don Diego, ... life is in danger!" says the friar.

e. The captain fights with .... sword.

f. The friars live in .... mission.

g. Lolita is angry and goes to .... mother.

5 Match the opposites.

1. wrong	**a.** disappear
2. old	**b.** alive
3. appear	**c.** dishonest
4. honest	**d.** fast
5. poor	**e.** right
6. slow	**f.** light
7. dead	**g.** young
8. heavy	**h.** rich

6 Who is your favourite character? ....

KEY TO THE EXERCISES AND EXIT TEST

KEY TO THE EXERCISES

CHAPTER 1

Page 14 Exercise 1
a. bandit
b. southern
c. takes, gives
d. friend
e. wears
f. travels
g. leaves

Page 14 Exercise 2
a. No, he isn't.
b. He is young and handsome. He has black hair and dark eyes.
c. He wears fine clothes.
d. He likes music and poetry.
e. Yes, because Zorro is sincere. He punishes only bad people. He protects the poor, the natives and the friars.
f. No, he is rich.

Page 15 Exercise 3
France – French
Italy – Italian
U.S.A. – American
Germany – German
Great Britain/England – English

a. He is Italian.
b. She is French.

c. They are American.
d. We are English.
e. You are German.

Page 16 Exercise 4

 stormy

 foggy

 windy

 sunny

 snowy

 cloudy

 rainy

a. It's foggy.
b. It's sunny.
c. It's snowy.
d. It's windy.
e. It's a rainy day.

CHAPTER 2
Page 20 Exercise 1
a. opens
b. laughs
c. looks
d. pulls
e. watches
f. jumps
g. slaps
h. makes
i. runs
j. jumps

Page 20 Exercise 2

S	ES	IES
travels	washes	studies
rains	goes	cries
plays	watches	
works	does	
starts		

a. washes
b. travels
c. rains
d. goes
e. studies
f. beats
g. watches

Page 21 Exercise 3
a. rich – poor
b. alive – dead
c. left – right
d. bad – good
e. heavy – light
f. fast – slow

Page 21 Exercise 4
b. It's (a) quarter past nine.
c. It's twenty-five past six.
d. It's half past eight.
e. It's twenty to two.
f. It's (a) quarter to four.
g. It's five to five.

CHAPTER 3
Page 25 Exercise 1
a. on, to   **b.** of   **c.** with   **d.** at
e. from   **f.** in   **g.** into, into

Page 25 Exercise 2
warm, horse, hacienda, friend,
family, rich, happy, surprise.

Page 26 Exercise 3
a. face
b. heart
c. hand

M	L	I	O	Y	F	D	U	N	Z	S
P	O	J	F	A	C	E	H	Q	I	K
H	G	V	N	W	S	E	Y	U	E	X
Z	O	M	C	G	A	P	B	R	M	O
D	H	G	E	R	S	I	L	C	N	O
T	I	M	T	H	A	N	D	S	Y	P
D	G	J	E	A	S	V	O	I	M	R

Page 26 Exercise 4
a. Lolita is Don Carlos' daughter.
b. She is 18 years old.
c. She has got long, black hair and
dark eyes.
d. She wants to marry a strong,
romantic man.
e. She loves Zorro because he is a
courageous man.

Page 27 Exercise 5
a. F – The next day Don Diego rides
to the Pulido hacienda.
b. F – The Governor doesn't like Don
Carlos.
c. T
d. T
e. F – Don Diego always gets tired.
f. F – Lolita is angry, so she runs
away.
g. T

SPANISH CALIFORNIA
Page 32 Exercise 1
a. warm and sunny
b. a mission
c. in the village near the mission
d. friendly
e. big haciendas
f. the Pacific Ocean
g. Spain
h. military presidios

CHAPTER 4

Page 37 Exercise 1
a. Don Carlos and his family have dinner at 8 p.m.
b. Don Carlos, his wife and daughter stand up.
c. A Spanish soldier enters the house and wants to arrest Zorro.
d. Zorro and Captain Ramón are good fighters.
e. Zorro injures the Captain's shoulder.
f. The Captain likes Lolita.
g. Don Diego sends a letter to Don Carlos.

Page 37 Exercise 2
a. the family; **b.** a servant; **c.** Lolita; **d.** Zorro; **e.** Captain Ramón; **f.** many soldiers; **g.** Don Diego.

Page 38 Exercise 3
a. his; **b.** his; **c.** her; **d.** their; **e.** your.

Page 38 Exercise 4
teach - teacher
sing - singer
dance - dancer
play - player
report - reporter
photograph - photographer
speak - speaker

CHAPTER 5

Page 42 Exercise 1
home, big, library, enters, hand, kiss, slaps, appears, villain, door, out, embrace

Page 42 Exercise 2
a. villain
b. flowers
c. door
d. library
e. garden
f. trees
g. hand
h. face

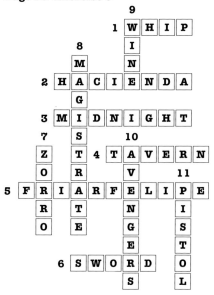

CHAPTER 6

Page 47 Exercise 1
a. He returns to the Presidio.
b. He writes a letter to the Governor.
c. His horse is very fast.
d. They are tired and angry.
e. He is a friar.
f. He returns to his father's hacienda.

Page 47 Exercise 2
people, morning, friar, chains, poor, magistrate, two, whip, friend, father

CHAPTER 7

Page 52 Exercise 1
a5 - b2 - c3 - d6 - e1 - f8 - g7 - h4.

Page 52 Exercise 2
a. dishonest; **b.** young; **c.** right; **d.** appears; **e.** together.

Page 53 Exercise 3

```
                              9
                      1  W H I P
              8          I
              M          N
      2  H A C I E N D A
              G
      3  M I D N I G H T
      7       S        10
      Z       T  4  T A V E R N
      O       R        V        11
   5  F R I A R F E L I P E
      R       T        N        I
      O       E        G        S
                       E        T
              6  S W O R D      O
                       S        L
```

CALIFORNIA TODAY
Page 56 Exercise 1
state, people, missions, speak, names, capital, natural, parks, young, universities, cosmopolitan

Page 56 Exercise 2
a. San Francisco, Los Angeles
b. Yosemite
c. Berkeley

CHAPTER 8

Page 60 Exercise 1
a. traitors
b. soldiers
c. honest
d. death
e. lake
f. mask
g. rescue
h. hacienda

Page 60 Exercise 2
Pass the word to everyone.

CHAPTER 9

Page 66 Exercise 1
a. F – The soldiers follow Zorro. They cannot capture him.
b. T
c. F – Zorro puts his pistol to Captain Ramón's head.
d. T
e. T
f. F – Zorro and Lolita hide inside the old tavern.
g. F – The Avengers come to rescue Zorro and Lolita.
h. T
i. F – Zorro takes off his black mask.

Page 66 Exercise 2
rescue, Governor, old, happy, face, black, surprised, Alejandro, Diego, Lolita

KEY TO EXIT TEST

1. 1. Spanish California 2. first 3. fox 4. friars 5. rich 6. big
2. a. F – Zorro wears a black mask and rides a fast horse.
   b. T
   c. T
   d. F – Don Diego thinks that marriage is boring.
   e. T
   f. T
   g. T
   h. F – Friar Felipe is an honest man.
   i. T
   j. F – Zorro kills Captain Ramon with his sword.
   k. F – The man behind the mask is Don Diego.
3. a. jumps b. goes c. run d. studies e. watches f. washes g. does
4. a. his b. her c. their d. your e. his f. their g. her
5. 1. e 2. g 3. a 4. c 5. h 6. d 7. b 8. f
6. Open answer.

NOTES

NOTES

NOTES

NOTES

Black Cat English Readers

BLACK CAT ENGLISH CLUB
Membership Application Form

BLACK CAT ENGLISH CLUB is for those who love English reading and seek for better English to share and learn with fun together.

Benefits offered:
- *Membership Card*
- *Member badge, poster, bookmark*
- *Book discount coupon*
- *Black Cat English Reward Scheme*
- *English learning e-forum*
- *Surprise gift and more...*

Simply fill out the application form below and fax it back to 2565 1113.

Join Now! It's FREE exclusively for readers who have purchased *Black Cat English Readers* !

The book(or book set) that you have purchased: _____

English Name:_____ (Surname) _____ (Given Name)

Chinese Name: _____

Address:_____

Tel: _____ Fax: _____

Email:_____
(Login password for e-forum will be sent to this email address.)

Sex: ❏ Male   ❏ Female

Education Background: ❏ Primary 1-3   ❏ Primary 4-6   ❏ Junior Secondary Education (F1-3)
❏ Senior Secondary Education (F4-5)   ❏ Matriculation
❏ College   ❏ University or above

Age: ❏ 6 - 9   ❏ 10 - 12   ❏ 13 - 15   ❏ 16 - 18   ❏ 19 - 24   ❏ 25 - 34
❏ 35 - 44   ❏ 45 - 54   ❏ 55 or above

Occupation: ❏ Student   ❏ Teacher   ❏ White Collar   ❏ Blue Collar
❏ Professional   ❏ Manager   ❏ Business Owner   ❏ Housewife
❏ Others (please specify: _____)

As a member, what would you like **BLACK CAT ENGLISH CLUB** to offer:

❏ Member gathering/ party   ❏ English class with native teacher   ❏ English competition
❏ Newsletter   ❏ Online sharing   ❏ Book fair
❏ Book discount   ❏ Others (please specify: _____)

Other suggestions to **BLACK CAT ENGLISH CLUB**:

_____

Please sign here: _____

(Date:_____)